TOXIC WASTE
AND RECYCLING

© Aladdin Books Ltd 1988

Designed and produced by
Aladdin Books Ltd
70 Old Compton Street
London W1

First published in the
United States in 1988 by
Gloucester Press
387 Park Avenue South
New York, NY 10016

ISBN 0 531 17080 2

Design: Rob Hillier
Editor: Margaret Fagan
Researcher: Cecilia Weston-Baker
Illustrator: Ron Hayward Associates

Library of Congress Catalog
Card Number: 8782 887

Printed in Belgium

The front cover photograph shows workers dressed in protective clothing handling toxic waste at sea.
The back cover shows a massive collection of broken glass ready for recycling.

The author, Nigel Hawkes, is diplomatic correspondent to The Observer *newspaper, London, and author of several books on nuclear power and energy issues.*

The consultant, Brian Price, is a freelance pollution consultant and a university lecturer in environmental health.

Contents

TOXIC WASTE
AND RECYCLING

NIGEL HAWKES

Illustrated by
Ron Hayward Associates

Gloucester Press
New York : London : Toronto : Sydney

Introduction

We live in a throw away society. Every year the waste increases, both in bulk and in complexity. The world chemical industry has increased its production, and its waste, by at least 15 times since 1945. A lot of the waste produced is simply unsightly but some can be toxic. Toxic waste is poisonous. It pollutes rivers and lakes, is dumped in the sea and can cause severe health problems. Some waste, such as highly active nuclear waste, is too dangerous even to dispose of.

But not all the problems are created by industry. Domestic rubbish on a dump can also be toxic especially when it contains items like old batteries or paint thinner.

▽ Rain water washes through the dump dissolving any toxic chemicals thrown out as domestic rubbish. The chemicals ooze from the bottom of the dump as a thick, poisonous sludge.

We talk of the "disposal" of waste, but this is really the wrong word. Waste can only be shifted from one place to another, or converted into a different form. Waste that is buried may be invisible, but it has not been disposed of. It may still cause problems if houses are built on it or if it should contaminate water supplies. Waste that is burned is turned into gas and smoke that can pollute the air. Much as we would like it, waste cannot be made to vanish. It has to be controlled, monitored, managed – or recycled.

Who should control the waste and who should pay for any accidents or clean up the pollution caused? These are the issues discussed in this book.

Industrial waste

Most industrial processes generate wastes. In the United States, almost half the hazardous wastes come from the chemical industry, a fifth from the extraction and purification of metals, an eighth from petroleum and coal products. Most of the waste is dumped in landfill sites — holes in the ground — or in pits, ponds, or lagoons. Sixty-five per cent of hazardous US wastes, and 85 per cent of the equivalent British wastes, are disposed of in this way. Many old landfill sites are a cocktail of dangerous chemicals, mixed together haphazardly with no official records kept. Cleaning them up would cost billions of dollars. In the United States this clean up is often financed by the tax payers' money.

▽ The manufacturing and chemical industries generate almost 100 million tons of waste a year in Britain alone. About 5 million tons is classified as "special waste" posing a threat to human life and health. However, many people consider the category "special waste" to be inadequate: some long-lasting and non-lethal health problems are not covered by the term and neither is environmental damage.

◁ Many of the most deadly wastes end up stored in drums, too toxic to throw away.

In many countries government schemes, such as those organized by the Environmental Protection Agency in the United States, encourage companies to dispose of waste safely. Aided by these schemes, the richer chemical industries now spend huge amounts of money on toxic waste control. Before the waste is dumped, it may be processed to make it less active and therefore less hazardous.

Some waste is recycled. The Minnesota Mining and Manufacturing Company, 3M, makes video tapes. As a byproduct they produce a toxic chemical – ammonium sulphate. They sell this to fertilizer makers who convert it to plant food.

◁ A petroleum refinery in California, run by Chevron, has found one solution to the toxic waste they produce from oil. They plow it into the soil. There the wastes are attacked by bacteria that live in the soil. The oily waste is converted into non-toxic carbon dioxide and water. Waste-eating bacteria are becoming more widely used. In Britain, waste-eating bacteria are being developed to clean up land which may have become contaminated due to an industrial accident or toxic waste dumping. The bacteria feed on the toxic chemicals.

Burning and dumping

One alternative to burying industrial waste in landfill sites is simply to store it in special containers. Another is incineration. However, this is an expensive process — it costs three times more than landfill. Moreover, burning chemical wastes is a difficult job. Unless the furnace temperature is very high, toxic gases can escape. Over the years, people living around chemical incineration plants have repeatedly complained of inexplicable diseases in animals and humans. The basis of these claims has never been scientifically established. Steam produced by some incinerators is used to provide heat for domestic housing. Some companies run their own incineration plant which reduces the risk of an accident in transit.

Landfill sites are now being phased out in the United States because they can "leak," pollute the ground water and contaminate the water supply. But in Britain, landfill is still widely accepted. In fact, Britain even imports hazardous wastes from the rest of Europe. There, safety standards are tighter and it is more expensive to get rid of waste.

▽ Out of sight of land, the incinerator ship *Vulcanus* drops anchor in the North Sea and burns potentially dangerous wastes. The sea is supposed to soak up any toxic materials in the smoke plume. But monitoring the ship is impossible, so it is likely to be banned. In 1987, the environmental group Greenpeace tried to stop the *Vulcanus* in protest at the burning of toxic waste at sea.

◁ Mining uranium from open cast mines gives off a gas called radon. If inhaled this gas can cause cancer. Mines exist in the United States, Canada, Australia and Namibia. Safety controls in the mines vary widely.

Worldwide controls on toxic waste and pollution vary. West German and Dutch wastes go to East Germany and Czechoslovakia where pollution controls are slack. Some people in the Third World claim their countries are being used as hazardous waste dumps for richer countries. Wastes from the United States are sent to Panama for dumping under appalling conditions. Safety standards in factories or mines producing toxic waste as a byproduct also vary. Many practices outlawed in the West continue elsewhere with ill effects on people's health.

▷ Stump Creek Gap in New York State was nicknamed the "Valley of the Drums" after 25,000 steel drums containing toxic waste were swept by flood waters from a dump into the valley. The US Environmental Protection Agency has begun the massive clear-up operation.

Poisoned communities

▽ In Seveso, northern Italy, a cloud of deadly vapor was released from a chemical plant after an accident on July 10, 1976. It contained small amounts of dioxin which blew across a wide area, killing pets and making people ill. The town had to be evacuated and millions of tons of soil were stripped away (below) in a huge clean-up operation.

The problems from hazardous wastes may take years to show themselves. In the 1930s, a chemical company near Niagara Falls dumped waste chemicals, stored in steel drums, in a muddy ditch. In the 1950s the ditch was filled in, a school was built on top and a community called Love Canal was created by speculative builders. In the late 1970s, the drums began to leak: trees turned black, there was an awful smell and the oozing slime burned holes in children's shoes. People complained of nervous disorders and liver problems, and miscarriages ran at 50 per cent higher than normal. Scientists found 82 different chemicals at the site, 11 of them suspected or known to cause cancer. Love Canal was declared a Federal emergency and the population was forced to leave.

Dioxin
Dioxin (full name tetra-chlorodibenzo-para-dioxin, or TCDD) is produced as a byproduct in the making of a chemical, trichlorophenol, used to make herbicides. It causes severe skin complaints (chloracne).

PCBs
PCBs (polychlorinated bi-phenyls) are chemicals found in plastics, paints, pesticides and refrigerators. They are also in our bodies, in small amounts, picked up from the air. They cause cancer in high doses.

Love Canal is not the only community destroyed by chemical waste. At Times Beach, in Missouri, used oil was sprayed on the dirt roads in the early 1970s to keep down dust. It was severely contaminated with dioxin, one of the most potent of chemical wastes. Ten years later, when the dioxin was discovered, all 2,200 inhabitants had to leave; it was impossible to clear it up.

The chemical compound PCB also has deadly effects. At Swartz Creek in Michigan State tons of PCB were dumped illegally in a pond, killing off its wildlife. PCB compounds are still dumped in the Rhine. The chemicals enter into the food chain. First they are eaten by fish and eventually end up in the bodies of seals which feed on the fish at the Dutch end of the Rhine.

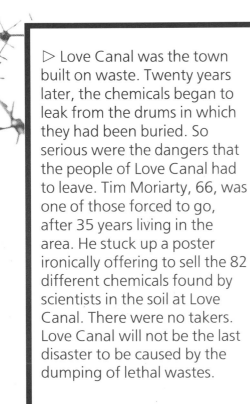

▷ Love Canal was the town built on waste. Twenty years later, the chemicals began to leak from the drums in which they had been buried. So serious were the dangers that the people of Love Canal had to leave. Tim Moriarty, 66, was one of those forced to go, after 35 years living in the area. He stuck up a poster ironically offering to sell the 82 different chemicals found by scientists in the soil at Love Canal. There were no takers. Love Canal will not be the last disaster to be caused by the dumping of lethal wastes.

Nuclear waste

△ The symbol above is used internationally to warn of the danger of exposure to nuclear materials.

▽ Sellafield, in Cumbria, is where Britain's nuclear waste is produced and stored. It is a reprocessing plant, where spent fuel from nuclear reactors is separated into its different ingredients, one of which is highly radioactive waste. There have been many minor leaks, and one major accident, at Sellafield, and local residents believe the plant has contributed to a locally high level of cancer deaths among children.

Few kinds of waste cause more controversy than the byproducts of nuclear power. As the uranium fuel is consumed inside a reactor, it produces a complex mixture of lethal materials. Nuclear wastes are radioactive – they produce invisible radiation which can injure or kill any form of life exposed to it. They retain their potency for hundreds, even thousands, of years, so they must be safely stored out of human contact in a depository that combines the sophistication of a laboratory with the security of a bank vault.

Because they are so dangerous, much more care has been taken with nuclear wastes than with other industrial wastes. But accidents have happened. The worst – never officially confirmed by the Soviet government – happened at a nuclear waste dump at Kyshtym in late 1957. An explosion spread radioactive materials over a wide area, contaminating it so severely that it is still unoccupied today. Much more minor accidents and leaks have occurred at Britain's nuclear waste plant at Sellafield in Cumbria, and at the American dump at Hanford Reservation in Washington State.

A radiation disaster happened in 1987 at Goiania, in Brazil, when a small phial of radioactive caesium from a disused hospital was broken open by a scrap merchant. His wife and daughter died. Here children in the area are checked for radioactivity. Brazil's Nuclear Energy Commission officially admitted that 243 people were contaminated.

▽ There are three categories of nuclear waste. Highly active wastes must be stored forever in secure conditions. Low-level and intermediate-level wastes can be dumped in properly engineered sites on land.

Accidents at nuclear power plants have drastic consequences too. In 1986, at Chernobyl in the Soviet Union, an explosion contaminated a large area with radioactive debris. Animals and crops had to be destroyed and topsoil removed. This soil is now a toxic waste and will be stored until the radioactivity decays to a safe level.

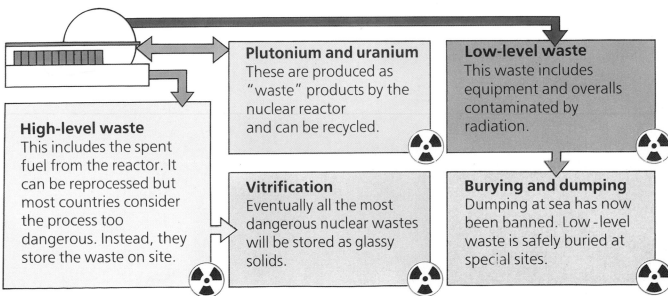

Plutonium and uranium
These are produced as "waste" products by the nuclear reactor and can be recycled.

Low-level waste
This waste includes equipment and overalls contaminated by radiation.

High-level waste
This includes the spent fuel from the reactor. It can be reprocessed but most countries consider the process too dangerous. Instead, they store the waste on site.

Vitrification
Eventually all the most dangerous nuclear wastes will be stored as glassy solids.

Burying and dumping
Dumping at sea has now been banned. Low-level waste is safely buried at special sites.

Compensation

Who compensates the victims of major waste catastrophes like Minamata? In principle, the polluter pays; but in practice this is not always true. It is often difficult to prove liability, or to demonstate negligence. If the number of victims is very large, paying compensation may force a company into bankruptcy. At Minamata the company paid token damages, $150 for relatives of adult victims and $45 for relatives of infant victims. Manville, an American asbestos company, faces claims of $112 billion for workers injured by asbestos. But many victims get no compensation.

$

Who pays?

Among the most dangerous of industrial wastes are heavy metals like lead, cadmium and mercury. In the 1950s, before the dangers had been realized, large amounts of mercury-containing waste from a chemical plant were allowed into Minamata Bay in southern Japan. The bacteria in the sea converted it into methyl mercury, which was concentrated in the bodies of fish. People who ate the fish soon began to show the classic symptoms of mercury poisoning – anxiety, irritability, followed by mental derangement and death. Thousands of people were affected; many of those who survived were reduced to a vegetable-like state.

Cadmium poisoning also had terrible results for one Japanese community. Haginoshima, on Japan's west coast, lay downstream from a mine whose waste was allowed to get into the river. During the 1960s the people there were exposed to dangerous amounts of cadmium, producing a horrible condition in which the calcium in the bones is replaced by cadmium, weakening them. The Japanese called the condition *Itai-itai* disease, after the cries of pain it produced from its victims – *itai* is the Japanese word for "ouch!" More than a hundred people died before the cadmium dumping was stopped in 1971. Many more survive, their bodies permanently fragile and broken by the poison they unwittingly swallowed in their diet.

Polluter pays

The law holds that industry is responsible for cleaning up its own wastes. And when offenders are caught polluting rivers or "fly-tipping" dangerous wastes, they are taken to court and fined. But such cases are a minority. The vast piles of dangerous wastes tipped into landfill sites over many years cannot now be traced to individual companies. They took care not to put their names and addresses on the drums before dumping them. So establishing legal responsibility is now impossible, and it will fall on governments to pay for the clean-up. The US Environmental Protection Agency's "Superfund," financed by taxes and contributions from industry, is organizing the clearance of thousands of dangerous sites. It will cost billions of dollars, and take many years to complete.

◁ A Minamata mother washes her daughter, crippled for life by mercury poisoning. Between 1956 and 1959, a third of the children born in the worst contaminated part of Minamata suffered from mental deficiencies; in total, as many as 10,000 people have been affected.

Toxic air

Among the air pollutants in cities is ozone, a form of oxygen. But while ozone at ground level may be a danger, high in the atmosphere it is a life-saver. A layer of ozone envelopes the earth and absorbs most of the Sun's ultraviolet rays, which would otherwise damage crops, alter climate and cause skin cancers.

They drift upward in the air until they reach the large "holes" which have been found in the ozone layer over the South Pole. The cause appears to be the propellants used in aerosols, chemicals called chloro-fluorocarbons, or CFCs. They drift upwards in the air until they reach the ozone layer, where they break down, producing chlorine which consumes the ozone. The gas is also found in refrigerator coolants and is emitted from fast-food containers. The damage caused by CFCs is so great that international efforts are now being made to replace them.

▽ This cyclist on the busy streets of Rome has sought protection for both himself and his child behind smog masks. A better answer would be to eliminate lead from gasoline. The United States introduced lead-free gas in the 1970s. With modification or redesign, car engines run perfectly well on it. The effects were swift: between 1976 and 1980, the lead in the blood of Americans fell by more than a third.

△ There is no money to be made in recycling old refrigerators. As they are left to rot on a dump, the CFC coolant leaks away into the atmosphere.

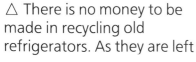

Another dangerous and unnecessary waste product is the lead that comes out of the exhausts of millions of cars. There is evidence that the lead in the air affects the brains of those who breathe it in. In today's cities, children forced to breathe traffic fumes have higher than average amounts of lead in their blood; and some studies show that these high levels are linked to lower intelligence.

Industry has also contributed to toxic air through pollution – particularly in the form of acid rain. The acid is formed when coal or oil containing sulfur is burned and produces, as a waste product, sulfur dioxide. This mixes with rain to produce the acid which can damage lakes, corrode buildings and kill trees. Acid rain is also formed from nitrogen oxides produced by power stations and cars. Coal power stations are now being designed to filter out the acid.

▷ The first alarms about acid rain were sounded by Sweden in 1972. Fish were dying as the acidity of its lakes and rivers rose. Later it became clear that trees were also suffering, particularly in West Germany. Acid rain has also struck the forests of North America. Now everybody is trying to cut pollution from power stations. As nitrogen oxides and ozone produced by car exhausts also contribute to the death of trees, stricter controls on car exhausts are called for. More than 20 countries are members of the "30 per cent club," pledged to cut sulfur emissions by 30 per cent by 1992.

Domestic waste

New York City is probably the world's garbage capital. It throws out 24,000 tons of domestic refuse a day, much of it packaging material that is often made from plastics. The rest of the developed world is not far behind. Where does it all go? The greatest bulk goes into landfill sites – but these are filling up fast and alternatives have to be found.

Nobody wants a dump in their backyard. Waste from cities is transported to the countryside, or even abroad, just to get rid of it. In March 1987 a huge barge carrying 3,100 tons of New York garbage sailed for a landfill site in North Carolina. But the site refused to accept it when New York officials refused to certify that it was not hazardous. So did sites in Alabama, Mississippi, Louisiana, Mexico and Belize. The barge and its tug, the *Break of Dawn*, spent months sailing up and down the Atlantic coast looking for somewhere to dump their cargo.

△ Litter is the most obvious rubbish problem in cities – and one of the hardest to control.

▽ The *Break of Dawn* sailed up and down the Atlantic coast looking for a place to offload New York City refuse, accompanied by millions and millions of flies. It finally finished back where it started when a landfill site in New York agreed to take its poisonous cargo.

How domestic waste has changed

Household refuse has changed dramatically in the past 50 years. There is less of it, by weight, because of the dramatic reduction of ashes and cinders from open fires. But the volume of garbage increased, with kitchen wastes, paper and plastics — many of them toxic when burned — all growing rapidly. Today ashes and cinders amount to only 13 per cent of the total, while paper has reached 30 per cent, plastics 7.5 per cent, and kitchen waste 26 per cent.

The throw away attitude extends to larger-scale items. Washing machines and cars are all dumped when they come to the end of their useful lives. In the past these were just left on dumps. Now there are more incentives to dispose of them properly. In some countries such as Norway, car-buyers pay a deposit when they buy the car. They get it back if the car is returned to a recovery center so that the metals can be reused.

Most drinks containers are now non-returnable. They end up being dumped as unsightly litter — a waste of both energy resources and materials. In the United States, the state of Oregon was the first to introduce a compulsory deposit on drinks containers to encourage reuse. Plastic objects are rarely reused. Consequently, a great volume of plastic is discarded every day which is a waste of valuable oil resources.

▽ Cars are crunched, parceled up and dumped in the minimum space possible. Every reusable metal has been removed. Nevertheless, unsightly dumps of scrapped cars disfigure the outskirts of many cities.

Living on the tip

Poor countries produce much less waste than rich ones: half as much per head in Calcutta than in London or Paris. And because they are poor they make sure nothing of value is wasted. In Cairo an entire community of 25,000 people, the Zabbaleen, live on what others throw out. Each morning they go out with their donkey-carts, collecting yesterday's rubbish. They concentrate on the richer areas, where pickings are better, but are paid nothing by the city. They earn only what they can make from the rubbish. Most of it goes to feed their pigs. Without them the city's refuse would lie on the streets uncollected, slowly turning into hazardous waste.

▽ Mothers and children from a Brazilian "favela" — shanty-town — search the local garbage heap. Refuse dumps are dangerous places, the more so as industry expands, often without adequate pollution checks to control the dumping of toxic wastes with domestic waste.

Similar scavenging systems operate in Mexico City. Ten thousand people survive on what they can find in the dumps. In Thailand, garbage collectors sort out paper, bottles, cans and plastics and earn a living from recycling. But there are grave dangers in such scavenging, as the accident with the nuclear source at Goiania, in Brazil, tragically demonstrated.

As industry expands, so do the risks of toxic waste. Although there are many poor people living in Brazil, parts of it are heavily industrialized. Petrochemical and steel plants in Sao Paulo, a huge industrial center, release wastes which have killed trees, birds and fish.

▽ New tourist hotels and offices in Lagos, Nigeria, back onto a garbage dump. Often new buildings like these are built without solving old problems first. Facilities for tourists may be adequate but the view from these hotel and office windows tells another story.

Recycling

The industrial world is also realizing that many waste materials are too valuable to throw away. Old cars are "cannibalized" for spare parts in scrapyards, then melted down as scrap which goes to make fresh steel for another generation of cars. Some glass bottles can be used up to 30 times if they are collected and cleaned. Broken glass – "cullet" – can be recycled to make new bottles, saving raw materials and energy. Waste paper can also be used again. In 1984, paper recycling programs in nine industrial countries spared a million acres of trees. Recycling aluminum drinks cans requires only a twentieth as much energy as making fresh aluminum from bauxite ore.

▽ Aluminum cans, used for drinks, can be recycled. At present such cans represent only about 1 per cent of British domestic refuse, but it is still worth trying to extract them because of the high value of the metal. New machines, which can sort out the cans from the rest of the rubbish by electromagnets, are being tested. Melted down (below), the old cans produce a new metal ingot.

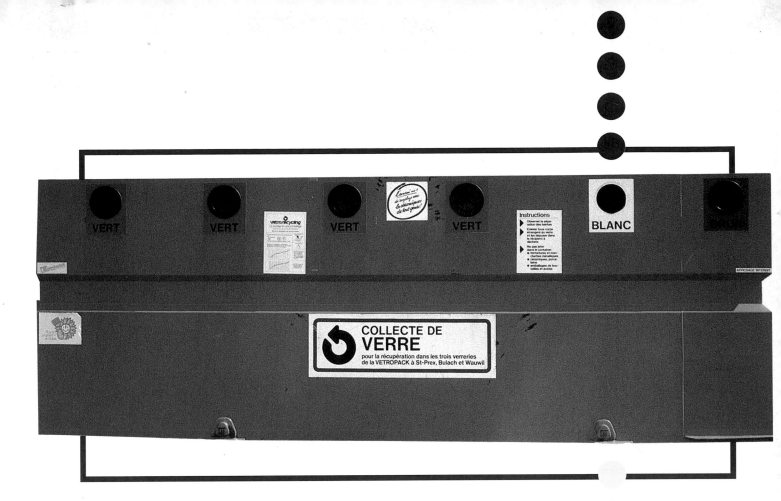

Despite this, only a fraction of domestic refuse is recycled. One major difficulty is that rubbish needs to be separated at source; paper, for example, rapidly loses value when mixed with other garbage. Organic materials – waste food and garden refuse – is best separated out for composting. Placed in large piles or special vessels to rot down, it turns into a valuable soil conditioner. In Sweden a quarter of all solid waste is turned into compost and recycled.

Efficiently separating the rubbish depends, to some extent, on public co-operation. In the United States, the public are encouraged to separate paper and glass for recycling and to avoid throwing away high-hazard items with the domestic waste. In the long term, recycling has many economic advantages. Not only does recycling use less energy than making things from scratch, it cuts down pollution by half. If items are designed with recycling in mind then the process is even more economical.

△ "Bottle Banks" have been set up throughout Europe. This one is in St Cergue in Switzerland. In Britain about 228,000 tons of glass is recycled every year, but only a small proportion comes from bottle banks. The bulk of it comes from industry, which is more systematic in the way it disposes of its waste. Other countries do better with bottle banks, but they depend ultimately on the public spiritedness of the users, who gain no benefit from the effort needed to sort their own rubbish.

After recycling

Once the "usable" rubbish has been separated, it still leaves the problem of what to do with the rest. Burning domestic waste has several obvious advantages. It leaves only harmless ash to be disposed of, and the furnace can be used to produce heat and electricity. But it has its drawbacks too like those associated with industrial incineration. The smoke produced may be unpleasant or even toxic, and it has until now been much more expensive than landfill. But despite the pollution risks and costs, many cities are turning to incineration in despair. In Denmark, Japan, Sweden and Switzerland, about half of household waste is now burned, producing steam which is used by industry or for heating nearby houses.

A better way to extract energy from waste may be to dump it in a landfill site and allow it to rot down. The process produces methane gas, which can be tapped off through pipes and used to provide fuel for factories or power stations. Worthwhile amounts of methane can be generated from ordinary rubbish, and landfill sites are now being designed with this in mind.

▷ Burning rubbish is more difficult than it appears, needing careful control if toxic waste products are to be avoided. The furnace shown here, at Ichikawa in Japan, is a simple incinerator designed to burn domestic waste. Some incinerators can produce so much heat and electricity that it can be sold to the national power supply.

▽ The photograph shows a dump in Britain which is still being infilled with refuse. The rubbish rots and produces the gas methane. A pipe beneath the dump leads the gas to the pumphouse which processes the gas. The gas is then supplied to a paper-making factory where it is used to fire a gas turbine which generates electricity.

Sewage and slurry

Sewage from humans and farm manure can cause enormous disposal problems. Where farm animals are kept indoors in large numbers, disposing of their manure can be almost impossible. Holland, which has 14 million pigs, is overflowing with pig manure. Dumped on the fields, it began to pollute canals and streams. The government set up "manure banks" but they were soon overflowing too. Now limits have been set on the amount of manure a farm can produce – forcing farmers to reduce their herds.

Manure does have uses other than as a fertilizer. As it rots, it produces methane gas and there are some attempts to use this. In China, for example, many villages use the gas – called biogas – to heat homes and provide light.

△ Sludge, or slurry, is produced in huge quantities from animal waste. Some of the manure can be spread on the fields as fertilizer. However, quantities produced are now beyond that which can be usefully recycled. Some is dumped at sea. Britain dumps 5 million tons of slurry a year. This helps to cause "enrichment." Plankton feeds on the excess nutrients and multiplies until it covers the sea's surface.

△ In western countries sewage plants, usually situated outside the major towns, process human waste until it is no longer a health risk.

▷ Sewage can be piped straight into the sea. Opinions vary as to the danger of this disposal method but it is disliked generally.

Human wastes that go down the drain are also a growing problem. Sewage is treated in treatment plants, and the sludge produced is used as a fertilizer, or dumped at sea. The water run-off from the plant is often allowed to run directly into the sea. As a result, many holiday beaches can be dangerously polluted.

In many Third World countries sanitation is inadequate. Untreated human waste often pollutes the drinking water which can lead to serious diseases such as cholera. The United Nations declared 1980-1990 the International Drinking Water Supply and Sanitation Decade. The aim is that richer countries will help poorer countries to improve their water supplies, sewage treatment and hygiene.

Not in my backyard

Many laws exist to control the disposal of rubbish and toxic waste, but enforcing them is difficult – especially on an international basis. Nobody wants waste in their backyard – whether it is unsightly litter or highly toxic chemical waste.

Britain, described as "the dirty man of Europe," is under criticism from other European countries on both sewage dumping and the incineration of toxic waste in the North Sea. The United States although cleaning up its landfill sites and introducing new toxic wastes controls still faces severe problems. Companies continue to defy the law and dump illegal and hazardous waste, often through city and county sewers.

OUR LIVES IN YOUR HANDS

NIREX

OUR COUNT OUR FUTUR

NO NUCLEAR DUMP

NO DUMP

RADWELL

NO NUCLEAR DUMP

WHERE IS

BOG OFF NIREX

NIREX NO

BRADWELL SITE
totally unsuitable
REFUSE Maldon MP · Gov't Chief Whip

ESSE NO D TAKE

NO NUCLEAR WASTE IN RADWELL

NUC WA RAD

I DONT WANT IT DONT COME BACK NIREX LEAVE US IN PEACE

▽ Nowhere is the "not in my backyard" attitude more apparent than in the case of nuclear dumping. Nirex, a British company, has provoked the anger of many local communities in its attempts to find a suitable landfill site for dumping low level waste. Residents voiced their lack of confidence in the safety measures being planned. There was a series of demonstrations throughout the country as the company looked for a suitable site.

To cope with the waste problem, factories are being forced to build recycling plants. And more sophisticated plans are being developed for the waste no one wants. Belgium plans to build the first "waste island" in the North Sea – although many fear it could be used for handling toxic and nuclear wastes without adequate safety controls and out of the public view. People are going to have to change their attitudes towards waste. Householders will have to separate their rubbish, to simplify and cut the cost of recycling. Industry will have to learn to dispose of its wastes properly and accept that there is a price to pay for the safe disposal of the waste so casually discarded.

▷ In many places in the United States, there are laws that prohibit littering. But since there are few people to make sure that the laws are obeyed, and these people are more concerned with criminal offences, not many of the laws are enforced. Some local governments even impose fines for fouling by pets and in some cities – notably New York – there have been strong attempts to keep streets clean. But this kind of cleanup is left largely to conscientious individuals.

Hard facts

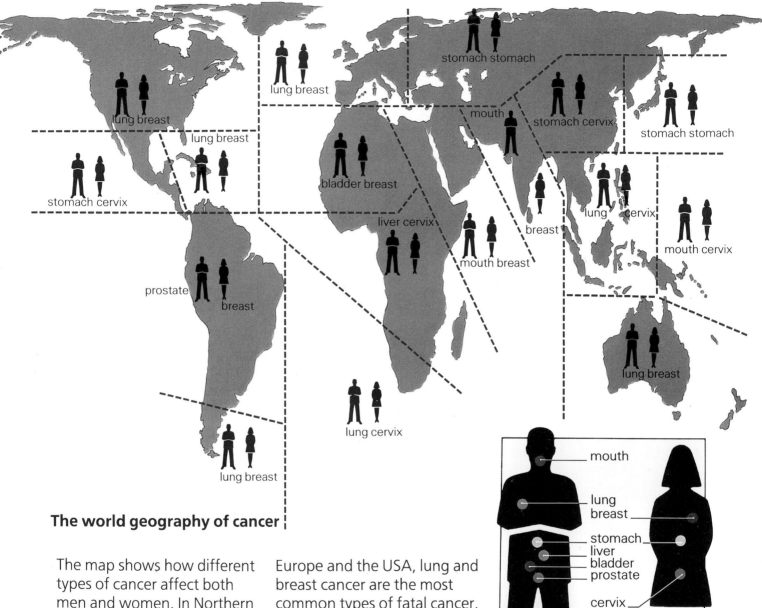

The world geography of cancer

The map shows how different types of cancer affect both men and women. In Northern Europe and the USA, lung and breast cancer are the most common types of fatal cancer.

The World Health Organization estimates that 60-70 per cent of all cancers are environmental in origin. So pollution and toxic waste may already be causing far more deaths than we yet realize.

Many toxic waste materials are carcinogenic – they can cause cancer. Often it takes years for the cancer to show itself which means that companies have sometimes been reluctant to accept a connection between illness and the work carried out by their employees. The situation is further complicated because many cancers are caused by personal habits such as smoking.

The world geography of cancer shows how the most common forms vary. There is little doubt that this variation is caused at least in part by differences in environment, including pollution – though diet also plays a major part.

Listed below are some of the countries with toxic waste problems and the solutions they have adopted. The problems affect both First World and Third World countries.

Brazil
This country is heavily industrialized and has major toxic waste problems. But as many of its factories have been built in recent years, some are equipped with the best pollution reduction technology. In fact, recycling technology is big business in Brazil.

Egypt
Like many Third World countries, Egypt simply lacks the resources to control the wastes of its rapidly growing population. The sewers of Cairo, designed for a city of one million people, now have to cope with seven million and are literally bursting.

Japan
As a highly industrialized nation, packed into a very small land area, Japan suffers badly from the consequences of pollution and toxic waste. But the disasters of the 1950s with mercury and cadmium poisoning have made the Japanese more aware of the dangers. The experience of the dropping of the atomic bomb on Hiroshima has also made the Japanese very cautious about nuclear power. This is in spite of their having no domestic oil reserves.

Poland
Heavily industrialized with old-fashioned steel works, Poland produces large amounts of toxic waste. The region around Katowice, near Cracow, spews 7 tons of cadmium, 170 tons of lead and 470 tons of zinc dust out of factory chimneys every year.

Sweden
Probably the most environmentally conscious nation on Earth, Sweden has done more than any other to clean up its countryside. It also has a well-thought out plan to bury all its nuclear waste at the bottom of a long tunnel running out to sea, where it believes it will be perfectly secure.

United Kingdom
As the first industrialized nation Britain has a long experience of pollution and waste. The Clean Air Act of 1956 banished forever the choking smogs which used to cause many premature deaths every winter, and rivers are cleaner than they once were. But Britain is a major exporter of acid rain to Scandinavia, has done little to control automobile pollution and has many dangerous landfill sites throughout the country which are toxic waste threats.

United States
The problems, like the US economy itself, are vast: thousands of chemical dumps lurking like unexploded bombs, polluted rivers, and coastal waters which once supported flourishing fisheries now almost bereft of valuable species like oyster and crab. Fortunately the USA is rich enough to be able to clean up the mess, if the will is there.

USSR
Hard facts about pollution and waste in the USSR are scanty. A vast country dedicated to industrial growth, it has put protection of the environment low on the agenda. Now reform is in the air: a paper plant that was polluting the world's largest body of fresh water, Lake Baikal, has been forced to close. But the world's worst nuclear accident, at Chernobyl in 1986, has polluted a large area of the Ukraine.

West Germany
The Germans today are concerned with the environment, awakened by the plight of their forests. Their priority is to control the acid rain which is killing the trees — but to do that they will need the help of other Europeans both east and west, who "export" their pollution. West Germany is also concerned with the severe pollution of the North Sea and the Rhine.

Index

Photographic Credits:

Cover and pages 6 (both), 10, 13, 16 (right), 17, 18 (bottom), 22 and 28: Rex Features; page 4-5: Photosource; page 7: Fred Ward/ Black Star/Colorific; pages 8-9, 20 and 21: Hutchison Library; page 8: Greenpeace; pages 9, 10 (inset) and 14: John Hillelson Agency; page 11: Popperfoto; page 12: Denis Thorpe/ The Guardian; pages 16 (left), 19 (left), 23, 27 (right) and 29: Robert Harding Library; pages 18 (top), 25 and back cover: Zefa; page 24: Davidsons Ltd, Purfleet; page 26: CIWF; page 27 (left): Water Authorities Association.